Revenue Engineering

Praise for *Revenue Engineering*

"While the notion of engineering your sales process has been around for decades, actually doing so remains rare. In *Revenue Engineering*, Desso-Cox not only lays out a straightforward, step-by-step guide, but also provides a robust set of practical implementation tools. Woe to the sales team who ignores this important work!"

—Todd Youngblood, Author
*Implementing Sales Process Media:
The Sales Professional's 24 X 7 X 365 Digital Assistant*

"The only thing that matters in business are your people – customers and colleagues. *Revenue Engineering* is for sales leaders that need to cut through the technical jargon while learning to use process and technology, and sound cultural values to bring a stellar customer experience into their organizations."

—Rick Fraumann, Director of Sales
Texas Disposal Systems, Inc.

Revenue Engineering

5 STEPS TO A CRM-READY SALES PROCESS

Tricia Desso-Cox

This is a work of fiction. Businesses, places, events, and incidents are either the products of the author's imagination or used in a fictitious manner. Any resemblance to actual businesses or events is purely coincidental.

Copyright © 2021 Tricia Desso-Cox. All rights reserved.

ISBN-13: 9780997680324

C5 Insight, Huntersville, North Carolina

Table of Contents

Introduction vii
   Why Process? vii
   Why Now? viii
   What to Expect ix
Foreword xi
About This Book xiii
The LUCK Principle™ xv
Fast Track xvii

Chapter 1 Listen 1
   Personas and Journey Maps 1
      The Challenge 2
      Toolbox 4
      Successful Application 20

Chapter 2 Understand 22
   Current and Future State Process Mapping 22
      The Challenge 23
      Toolbox 25
      Successful Application 35

Chapter 3 Connect · 37
    Technology Blue Printing · 37
        The Challenge · 38
        Toolbox · 40
        Successful Application · 48

Chapter 4 Know · 49
    Peer Accountability and Process Improvement · · · · · · · · · 49
        The Challenge · 50
        Toolbox · 52
        Successful Application · 61

Chapter 5 Good LUCK · 63
    Culture and Change Management Includes
        Leadership Engagement · 63
        The Challenge · 64
        Toolbox · 66
        Successful Application · 79

Conclusion · 81
    Revisit Annually · 81
    Tackle Other Processes · 81
    LUCK Resources · 82
References · 83

Introduction

Why Process?

The sales teams of today are facing the toughest buyer audience in the history of selling. Buyers are able to self-serve by gathering a wealth of information about a company and its products or services before they even have to think about reaching out to a salesperson for assistance. But when that prospect or customer does reach out, they're expecting a wow experience. From start to finish, your organization needs to already understand what your prospect actually wants to talk about at the beginning, middle, and end of the buying journey and throughout the longer customer life cycle.

Did you know that salespeople spend just one-third of their day actually talking to prospects? On average, they spend 21 percent of their day writing emails, 17 percent entering data into various tools, another 17 percent prospecting and researching leads, 12 percent going to internal meetings, and 12 percent scheduling calls. Given the precious few hours a day these folks actually get to sell, let's take a few moments to consider what a lack of structure might do to their day.

Consider your own day for just a moment . . . How many items are on your to-do list? How do you know what order to do them in?

How many times are you interrupted by leadership, your peers, your direct reports, and your customers? You know as well as many others that if you don't plan to run your day, your day will run you. So, apply that premise to a rookie salesperson who may still be learning the ropes of your industry, your products, and your technology and who also may not be as disciplined with how they approach the day. Or the seasoned salesperson who has gotten comfortable with methods that may have worked a decade ago but isn't keeping up with new approaches to reaching customers or with new technology. Scary, right?

This book is packed full of insight into how to design a sales process that is better suited to maximize a salesperson's day, engaging them in conversations not only with the leadership team but also with peers and ultimately better serving the prospect and customer.

Why Now?

Based on a study by *Salesforce.com*, over the next three years, sales teams anticipate that guided selling and coaching capabilities will grow by 98 percent and that high-performing sales teams are 2.3 times more likely than underperforming teams to use guided selling. But you have to be ready for the effort required to implement a sales process across the entire organization. In some cases, you will be looking at a business process change, some technology change, the human capacity for change, and possibly some change in personnel as a result of all the other changes. Therefore, make sure your team is ready!

A recent *HubSpot Research* survey of salespeople revealed that more than half rely on their peers to get tips for improving: 44 percent looked to their manager, 35 percent to team training resources, and 24 percent to media. Therefore, in order to have an early impact on your team's ability to accept the change and its overall adoption, plan to engage the entire team in the process of designing the process.

This book will share ways to get the team engaged early, keep it engaged throughout the implementation process, and empower it to deploy the process successfully as a team.

What to Expect

There are two different approaches you can take when implementing a new sales process: a very detailed, thorough examination using a lot of different tools leaving no stone unturned, or an accelerated approach that examines just the basics of process design.

This book provides you with a significant amount of detail, should you choose to invest time in researching customer journeys and preparing customer persona profiles and salesperson profiles, then fully exploring your current state, outlining its challenges, and then fully outlining a desired future state based on a combination of those things. This could take anywhere between a few months to a full year, depending on how detailed you decide to get with each task.

However, not all organizations have the time or budget for that level of undertaking. In that case, the book has also designated which items are "must do" in order to fast-track your implementation process. Just look for the shamrock (✤) throughout the book to identify these items and fast-track your way through the design and implementation process.

Given that you and your team already have day jobs, you need to decide how to divvy the tasks and how much time is acceptable for each person to allocate to this project. There is a multitude of ways this can be achieved. The project can be handled internally or with the support of an external consultant. At a minimum, your internal team needs to provide the business strategy, institutional knowledge, and subject matter expertise required to map out your process. An external consultant can bring the experience of having worked with dozens of clients to help you avoid pitfalls and minimize the time required from your team. The right partner will combine both business consultation and technical implementation expertise.

Lastly, this initiative will not be an easy journey. You can expect an initial level of pushback from some or all of your sales team. In many cases, these individuals believe that their personal process is the right process, especially if they are high performers. As mentioned above, the earlier you can engage them in the ideation process, the better off you'll be.

The day-to-day work will get in the way. There will be disagreements on what the right stages, steps, and tasks are in the future state process. There will be uncertainty as you're working through the unknowns of new technology implementation. And there will be frustration as you roll out the process and the team feels like it's having to go slower as it learns before it can ultimately speed up. Hang in there. As Thomas Edison said, "I never did anything worth doing entirely by accident. They were achieved by having trained myself to endure and tolerate hard work."

Foreword

Success in sales leadership in the 21st century requires a new way of thinking and a new set of tools. Effective sales managers must be digital natives—as comfortable with technology and analytics as they are with managing a team and engaging with demanding prospects.

Just as the best athletes surround themselves with coaches, the best sales managers work with advisors and experts. These experts bring deeper skills in the areas of process design, project management, and technology. They understand and appreciate the persona of the sales manager and the sales rep. They've worked across marketing, sales, and customer care to deliver great customer experiences. And they know how to harness the power of digital workplace tools to create processes that are repeatable, improvable, measurable, and scalable.

Tricia Desso-Cox is that person. For years, I have had the pleasure of watching Tricia engage with dozens of the best sales minds to revitalize their processes, drive adoption of their CRM solutions, and engage their teams in improving the bottom-line. If you want to develop a sales process that your people will adopt and that will make a measurable difference to your bottom-line, Tricia is one of the best. Her many years of experience include six-sigma process assessment and design, customer relationship management (CRM) implementation, sales operations, project management, and sales consultation.

In this book, Tricia shares the secrets she has learned by working alongside some of the best sales leaders in the business. It's a lot of information in a very small amount of space, so have your notepad ready as Tricia takes you on a journey Powered by LUCK with better sales process design!

Geoff Ables
Author – *The Luck Principle*

About This Book

Each chapter contains several elements. First, you'll be greeted with a couple of quotes that speak to the general topic of the chapter. Next, you'll get some details related to a case study challenge that we will solve together in the chapter. The meat of the chapter will focus on the toolbox where the process and tools recommended to solve the challenge are presented as well as other best practices. Finally, we'll conclude the chapter with the case study's application success story.

The case studies are real-world examples tackled and solved by employing C5 Insight over the last decade. However, the organization's names have been changed to protect their anonymity and specific business processes and challenges.

The LUCK Principle™

The LUCK Principle is a framework at the heart of every relationship that empowers better collaboration. If you are developing a relationship with colleagues, customers, in your community, or with your kids—you're already practicing LUCK. When implemented correctly, appropriate technologies are a "force multiplier" that can enable an organization to practice LUCK across hundreds, thousands, or even millions of customer and coworker relationships.

The LUCK Principle was developed and refined over a 15-year period by C5 Insight. During that time, the team observed successful relationship building and collaboration habits at an interpersonal, meeting, and organizational level across hundreds of clients and case studies. Those who are successful excel in the four core areas of LUCK described below, whereas others performed poorly in one or more of the four areas.

Listen	Understand	Connect	Know
Capture and remember information.	Identify opportunities to deliver unique value.	Interact in more relevant ways.	Evaluate and continually improve the approach.

For C5 Insight and its clients, improved LUCK is not only what is delivered to its clients and their clients; it is also a framework for managing relationships, setting goals, and running projects. C5 Insight is passionate about empowering people to work together better, and LUCK is the methodology for delivering on that passion. C5 Insight delivers customer and employee process design and technology road maps. This includes a comprehensive LUCK Scorecard and Plan, including benchmarking surveys, cultural analysis, process design, and technology roadmaps.

Throughout this book, you will be immersed in the LUCK framework as you learn how to listen to your customers and your team, determine how to better understand the needs of each, create a blueprint that connects a robust process with people and technology, and develop a plan that measures both the team and the process.

Lastly, you'll learn how to focus on engaging leadership to ensure your organization is rooted in a culture of vulnerable transparency and continuous improvement.

Read on to learn how to develop a sales process that's Powered by LUCK.

Fast Track ☘

Identification and Prioritization Workshop	5
Persona Profile Sheets	10
Current State	25
Pain Point Identification and Ranking	26
Envisioning the Future State	31
Fit/Gap Analysis	43
Technology Blueprinting	45
Determine the Metrics	55
Trickle Training	71
Culture and Leadership Engagement	74

CHAPTER 1
Listen

Personas and Journey Maps

> "A good listener is not only popular everywhere,
> but after a while, he knows something."
> — Mark Twain

> "How you gather, manage, and use information
> will determine whether you win or lose."
> — Bill Gates

The Challenge
Client Information:

- Manufacturer of small equipment
- >100 Employees
- Existing CRM solution in place

Machine Manufacturer (MM) has been in business for nearly five decades. It is accustomed to selling its equipment via traditional outside sales reps. This approach has worked well so far, because the reps could conduct in person prospect and customer visits, and the equipment was small enough that it could be carried onsite to conduct demonstrations.

In just the past few years, Machine Manufacturer decided to expand its footprint by pursuing sales in the accessory business related to the small equipment and the industry that its customers operated in. However, it quickly found that the role responsible for buying the accessories was not the same role responsible for buying the machinery. Thus, trying to sell via the outside sales channel was proving to be ineffective.

As a result, Machine Manufacturer created a small inside sales team deemed responsible for calling on the accessory buyers. The team had roughly outlined some basic processes for the new resources and had trained them on the usage of the CRM platform. The team got to work, and while experiencing some success, MM felt that the progress in making an initial sale wasn't quick enough, and the ability to turn an initial sale into a repeat buyer was proving to be challenging.

So, Machine Manufacturer attempted to assess the work being done by the inside sales team and the tools the team was using to conduct the work. What it quickly found was that while information was being tracked within CRM related to the activities the team was conducting, there was no easy way to determine how much time individuals were spending on those activities and whether or not those activities were really having any impact on buyer behaviors.

Machine Manufacturer was unsure if the process was correct or if the tools the team was using were the right tools. The team was also unsure whether the scripts the inside sales reps were walking through contained the right messaging or whether they were even targeting the right audience. The team had no idea who the accessory buyer really was, and it needed some help to get the inside sales team on the right track.

Toolbox
Personas

In order to truly understand a prospect or customer, you must first spend a great deal of time listening. Some of the listening is literal—it happens via phone calls, emails, and face to face meetings. However, you can do a great deal of listening by collecting additional bits of information about your prospects and customers. One particular method revolves around the usage of personas. Personas are the characterization of a prospect or customer who represents a segment of your target market. They include demographic details like gender, generation, position or job title, and physical location. They also contain psychosocial aspects, such as who or what can influence the prospect, what competes for their attention, things they might think or feel, or what might cause them to become disengaged.

Getting a clear definition on the variety of personas the sales team may interact with is important to ensure your team is approaching that particular persona in the most engaging, relevant, and effective manner. Trying to reach a more seasoned buyer who prefers face-to-face meetings or phone calls via a method like email or text instead is likely to lead to a closed door. More and more often, salespeople are facing committee-style purchasing, where more than one persona may come into play when it comes to influencing and decision making. A recent *HubSpot* study showed that while 64 percent of the C-suite has final sign-off on a pending deal, 81 percent of employees not in the C-suite do influence purchasing decisions.

Let's look at a series of tools we used with Machine Manufacturer to get started with not only identifying the various personas but also diving deeper to really get to know them.

Identification and Prioritization Workshop

The first thing you'll need to do is determine who the personas are that your team is typically engaging with, and you'll need to decide which of those personas are the most important to learn about more deeply. In order to do so, we suggest conducting a workshop with both your sales and marketing teams together. It is important to have marketing engaged in the conversation as well, because it will ideally be supporting the sales team with the targeted company or product-related content as well as with lead-generation and marketing qualification efforts.

Once you have the teams gathered, take a few moments to explain to them why you've brought them together and also what the concept of a persona is. Next, ask them to call out the various types of personas they think are relevant to your market, industry, product, or service and to your client's market, industry, product, or service, since there may be differences in the roles that align with one another. Some examples from our Machine Manufacturer included: buyer, business owner, gate keeper, central purchasing, repair center technician, and retail store manager.

Finally, you're going to work together as a team to prioritize the personas you gathered. The intent is to focus on developing the profiles for the most important personas first. However, if you have a limited amount of time or budget, you can choose the top one to three to focus on. The easiest way to prioritize them is to conduct a voting session. This allows folks to express an opinion without biasing others or being biased by them. The first round of voting should be silent. Then the team can have some time to debate the results and influence one another before conducting a revote to settle on the final ranking.

Persona	Vote 1	Vote 2
Buyer	10	16
Business Owner	3	8
Gate Keeper	4	7
Central Purchasing	1	4
Repair Center Technician	2	2
Retail Store Manager	1	3

Empathy Maps

Empathy maps are just one of the many tools used to begin to delve into a more meaningful definition of a persona. It is designed to help you visualize customer information so that you can better consider what the customer really wants—not what you think they want! An empathy map is divided into several sections meant to guide you through the exploration of a persona's emotional aspects.

Positive Experiences: In the current situation, what kinds of experiences are mostly positive for this individual?

Negative Experiences: In the current situation, what kinds of experiences are mostly negative for this individual?

Pain Points: What pain is the individual encountering even in positive experiences? What pain is the individual encountering that is contributing to negative experiences?

Goals: What is the ultimate goal? What are they trying to achieve?

Tasks and Actions: What actions is this individual taking in order to achieve the goal?

Influences: Who (people), what (systems) and where (places—real or virtual) is influencing how this person acts, the results they attain, and the overall experience? What is the context in which they're experiencing this journey (what is stressing them, what other priorities are they dealing with)? How is the individual connecting and communicating with these (in both directions)? Consider drawing first and second tier influencers.

Thoughts, Feelings, Quotes: What kinds of things is this person thinking, feeling, or saying throughout the journey? What stresses, frustrates, or angers them? What delights them? What really matters to them?

Try to rephrase all of this insight into a few simple quotes, something like this: "I am _____, so _____ is important to me."

It is likely not necessary to make an empathy map for all of the personas in the list. However, you may choose to focus on this tool for the top one or two personas your team interacts with most frequently.

Use the template below (download at: https://gotluck.link/Revenue-Engineering) to get started.

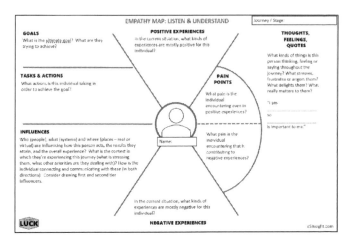

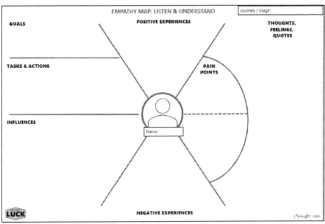

Influencer Diagramming

Drawing an influencer diagram can help you to think about situations where different personas influence each other in various ways and where you would perhaps like to influence the situation yourself. Think about a situation where your sales team is up against committee purchasing. It would be nice if, going into that situation, your team had a basic understanding of the dynamics of the relationships between the key roles on the committee. By taking some time to explore this concept, your team is better positioned to help ease friction, facilitate conversations, preempt a line of questioning that is likely to occur, and proactively provide information each role is going to need to convince the others to move forward with the purchase.

It is likely not necessary to do an influencer diagram for all of the personas in the list. However, if your team frequently deals with committee-style purchasing, you may choose to focus on this tool for the top one or two personas your team interacts with most frequently.

Use the template below (download at: https://gotluck.link/Revenue-Engineering) to get started.

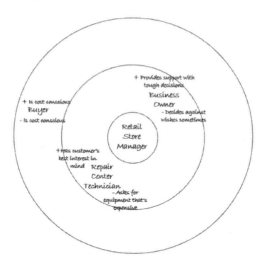

To generate an influencer diagram:

- Start with a large blank piece of paper.
- In the center, draw a small circle and write the name of the primary persona.

- Draw a larger circle around the primary persona circle.
- Now begin to think about the first degree of personas the primary persona may be interacting with during the buying cycle.
- Write their names within that second circle.
- For each one, think about a way that your primary persona may be positively influenced by this first-degree persona and a negative way they could be influenced. Write those beneath the first-degree persona, designating each with a + or a − symbol.
- Draw an additional larger circle around the first-degree persona ring.
- Repeat the same process with this layer, considering the second-degree personas that may be influencing both the first-degree personas and the primary persona you started with.

Persona Profile Sheets

A persona profile sheet is the aggregation of all of the information you've gathered using the tools we've discussed thus far. It is the final resting place for all of the data you've gathered about each persona that the team can refer back to as needed. If at any point moving forward the team needs to ask, "What would [persona] do, say, or feel," the persona profile sheet is the tool to provide the necessary gut check. The profile breaks down into four major categories: Listen, Understand, Connect, and Know. Each category contains several bits of data—some that you've already collected, some that you'll need to finish fleshing out.

> **Listen** is meant to explore how you capture data about this persona. Some of it will be demographic in nature, and some of it will be based on the historical data you may have about

them, if they are an existing customer, or perhaps a look-alike customer you're comparing to a prospect persona.

Understand begins to pull information from the empathy map, delving into the thoughts and feelings and things this persona might do or say.

Connect gets you thinking about how your company, product, or service and your business processes will entice this persona to do business with you. Key findings from the influencer diagram can be used here to think about messaging.

Know challenges you to think about continuous improvement in order to delight this persona not only with the first interaction but to turn them into a loyal customer who makes repeat purchases or expands their relationship with you.

We highly recommend putting together a persona profile for at least the top three to five personas that represent the top 80 percent of buyers your organization is going to encounter.

Use the template below (download at: https://gotluck.link/Revenue-Engineering) to get started.

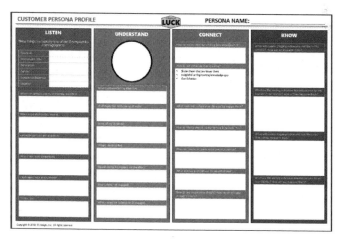

Exclusionary Personas

Exclusionary personas are exactly what they sound like—folks you want to weed out of the prospect qualification process as quickly as possible. We recommend the time-proven BANT (Budget, Access to power, Need, Time to decision) criteria to outline who these personas might be. Again, it may not be necessary to outline all of the exclusionary personas your team might encounter.

However, if there are a handful that you see time and time again, and it may be helpful to have these front and center in order to quickly disqualify them and move on. Most sales reps are optimists by nature, so a helpful tip sheet on disqualification can save a lot of time.

Use the template below (download at: https://gotluck.link/Revenue-Engineering) to get started.

Customer vs. Employee Personas

All of the tools discussed thus far have been put to use to further define, characterize, and segment your prospects. But what about your sales reps, yourself, and your leadership team? Each of those groups have individual needs, priorities, and goals as well. This can be especially true if your team is geographically dispersed, varying in age, divvied up by industry or solution, or just divided into those who may be tech savvy and those who aren't.

While you certainly can use each of the tools we covered to assess your team, we recommend at least completing the Identification and Prioritization workshop as well as the Persona Profile sheets. Even if you simply categorize your team as Sales Reps, Managers, and Executives, outlining the differences between those will help you distinguish business process needs, technology needs, communication preferences, training needs, and the desired level of access to information via reports, dashboards, and other analytical tools. The Persona Profile for your internal resources does vary a bit from the customer profile.

Use the template below (download at: https://gotluck.link/Revenue-Engineering) to get started.

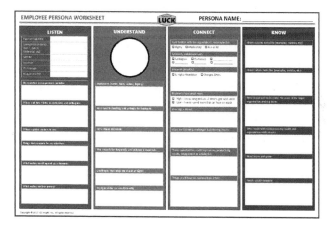

Journey Maps

A customer's journey is the sum of all the experiences they have while interacting with a company or brand. Most of the time, it starts before you even hear from them! They're gathering information via your website, what they read in industry publications, web reviews, their interactions on LinkedIn or other social media platforms, or what they've heard from their peers. Sales and marketing both play important roles in what the customer hears through these "pre-engagement" channels. However, once the prospect has raised their hand to start a conversation, it is critical to deliver a stellar experience from the moment interaction begins and to carry that through the entire buying process, culminating in a seamless handoff to any post sales activities that may occur.

A customer journey map is a visual representation of the interactions your prospects and customers experience before, during, and after the sales portion of their engagement with your company. It can include everything from searching for your product or service via the internet, to having a phone conversation with your sales rep, to getting a thank you email after completing a purchase, to submitting a complaint to your customer care team or having an item repaired via a technical service team, to making a repeat purchase and referring your brand to a colleague. It is a vast journey—broader than just the few moments in time that your sales team is engaged. However, to develop the sales process, you don't need to focus on the details of the entire journey. You can start by examining inputs to the sales process and carrying that through to any outputs of the sales process to another team.

A journey map can include multiple stages of a single process—in this case, the sales process. Alternately, a journey map can be completed for each individual stage of the process. The latter will be much more granular in nature, delving into every little activity, message, input, and output. We recommend starting with the higher-level multi-stage map to begin with. As the team implements the process and begins to learn more, a deeper dive map per stage may be more applicable. In either case, a journey map is divided into various segments in order to apply focus to both the tangible things that have to happen but also to the intangible things like the way the customer might be feeling about those interactions.

Goals

Goals are the foundation of the map. They are meant to identify not only what the customer needs or wants to get from that stage in the journey but also what your company needs or wants to get. If you don't have clear definition of what the goals are, you don't have clear definition of the stage the customer is in.

Transactions and Interactions

Transactions and interactions are the listening mechanism. This is where you'll capture the persona's needs, applicable data from your persona profiles that might be helpful, questions the persona might have for you, or conversely questions your team should have for them, how the persona might be feeling, what they might be saying, and things that they or your team might be doing.

Experiences

The Experiences segment is where you will rate the interactions that have been outlined. Meets Expectations is the status quo and is the minimum viable experience you're aiming for. Below Expectations is typically synonymous with aspects of the process that are painful to execute or where the customer is feeling an aspect of disappointment. Exceeds Expectations is the wow factor. This is what you're aiming for, not necessarily for every single interaction but at least once in every stage of the overall sales journey.

Influences

Influences are the things that manage or impact the interactions the customer is engaged in. They're divided into two different groups—the front office communication channels and people, and the back-office business processes and technology. There are a lot of details to flesh out related to this segment of the journey map. However, the minute details will not reside on the map itself; they will be outlined in your current and future state process maps.

Opportunities to Improve

The Opportunities segment is where you will begin to brainstorm ideas to improve the Below Expectations experiences you previously identified. You will use a standard difficulty/impact quadrant matrix to categorize each idea to assess the estimated level of difficulty and potential impact the idea could have on resolving the negative experience.

Journey Mapping Workshop

To execute a journey map workshop, gather your sales team or at least a sampling of different levels of experience both from a selling perspective and from a technology perspective. The most important thing to remember while working through this exercise is that you should be focusing on what the customer is experiencing, not what your team is experiencing. Be certain you're working from the viewpoint of the customer and remind yourselves often to do so.

Use the template depicted above (download at: https://gotluck.link/Revenue-Engineering) to get started.

Persona Review and Selection

- Start by reviewing the personas that were developed.
- You'll need to choose a persona to focus on while building your first journey map. There are a few ways to think through this:
    o Use the 80/20 rule to select the persona that most broadly fits the definition of your product or service's typical buyer.
    o Or think of it as the typical buyer for the product or service you sell the most of.
    o Or choose the typical buyer for the product or service you want to grow if that's the goal of designing a new sales process.
- Once you've selected the persona, put a copy of their persona profile sheet on the left side of the journey map to refer to as needed.

Current State Customer Journey

- Start by labeling the name of the journey the team will be focusing on in the Journey Name spot on the map.
- Next, list the high-level stages a prospect or customer engages in while working with your sales team across the top of the map, just above the goals section. Machine Manufacturer had five stages a customer went through. If you're unsure you can simply list some generic terms, like Information Gathering, Quoting, Decision Making, Purchasing, and Supporting (or even more basic terms like Beginning, Middle, and End) to provide some general time designations.
- Next, use small post-it notes to jot down the goals for both the customer and the business in each of those stages and place them in the appropriate boxes.

- Continue working through the map from top to bottom, using small post-its to capture the appropriate information for each item. Place the post-its in the appropriate boxes. As you are working through, you may capture information that causes you to question something that you previously captured; that's ok. Explore all of the thoughts and ideas in order to capture as much information as possible. Keep in mind that a journey map is not a flowchart; you can include items that only happen sometimes (i.e., "customer portal is down").

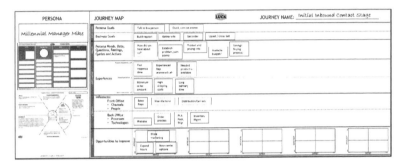

Future State Customer Journey

- The future state journey map is built in the same manner as the current state; however, you want to focus on incorporating the ideas for improvement where applicable in order to develop a state-of-the-art "wow factor" journey. The question you're trying to answer is, "If we could deliver a stellar buying experience for our customer, what would that look like?"
- Remember to review all of the items in the Experiences section that were Below Expectations or Meets Expectations to see if you can increase satisfaction in any of those areas. Don't be afraid to challenge yourselves as a team and an organization to make sure that you're at a minimum getting back to

the basic etiquettes of conducting business but to also think outside the box in how you're doing that or to innovate exciting new ways to engage that your competitors aren't yet doing. What Uber and Lyft did to the personal transportation industry is a prime example of conducting business, with basic business etiquette via an out of the box, exciting new way that its peers—the existing taxi companies—weren't even thinking about.

The future state customer journey will help you to identify what is missing in your internal current state sales process that is necessary to fulfill the desired future state.

Successful Application

We last left our client, Machine Manufacturer, with a challenge on its hands. The team needed some assistance identifying what an accessory buyer looked like and how to best interact with them. As a result, MM participated in a three-day workshop that engaged its outside sales, inside sales, and marketing teams.

The key workshop activities included:

Begin with the End in Mind: The team spent time discussing team key performance indicators (KPI) that the business needed and wanted to track and how to capture them. Then those items were prioritized in order to determine which ones should be displayed on dashboards and in what layout. In order to measure these items, the team had to design the future state, making sure information about each KPI was being collected throughout the process.

Ride Along: By observing the inside sales team conduct its day-to-day work, the project team was able to identify differences in how the current state process was executed from rep to rep. The team was also able to assess gaps in technology, as the reps were not able to find all of the information they needed within just one or two systems, and the process instead required accessing upward of five different applications on a daily basis. The ride along also identified pain points throughout the process where the user had to back track, enter duplicate information, search for information from multiple sources or worst case, have to call a customer back because they couldn't complete the interaction.

Personas: The team ended up identifying seven major personas and completed Persona Profiles for three of them. The marketing team was able to provide valuable insight to the profile generation based upon those personas' prior interactions with the company website and other marketing collateral. Outside sales was able to speak to the Exclusionary Personas, since they had experienced the lack of success with the equipment buyers they had initially tried to sell accessories to. Inside sales had a great deal of interaction information to contribute based upon the trial and error it had been through already.

Journey Mapping: Once the team had completed the persona profiles, it chose the most common persona to focus on to develop "the 80 percent happy path process" that a typical accessories customer would encounter. The marketing team stayed engaged in developing the journey so that it could identify where additional outbound messaging and content marketing would be helpful and aid in crafting call scripts for the inside sales reps and templates to share additional information with the customer.

Via the workshop, the team defined five customer journey stages that a customer was engaged in from initial contact through post purchase advisory and support work. The team also identified 15 opportunities to improve the experience the customer was having via its transactions and influences throughout the process. In the end, the inside sales team was able to compare and contrast methods to come up with a cohesive, best-practice path forward based upon what it had learned about the personas, the stages, and the identified improvements that the entire team agreed to execute against.

CHAPTER 2
Understand

Current and Future State Process Mapping

> "If you can't describe what you're doing as a process, you don't know what you're doing."
> — W. Edwards Deming

> "Learning is a process where knowledge is presented to us, then shaped through understanding, discussion and reflection."
> — Paulo Freire

The Challenge
Client Information:

- Global Beverage Solutions
- >500 Employees
- No CRM solution in place

Global Beverage Solutions (GBS) is a fairly young organization, having only been in business for two decades to date. However, the organization grew fairly quickly into a global entity spanning six continents in over 50 countries. Sales were managed via a team-based model incorporating sales reps, design engineers, and distributor sales reps. As a result, tracking the pipeline from lead all the way through opportunity and on to repeat business proved to be challenging.

The company was also tracking contact information, related communications, and budgetary estimates in disparate tools like Outlook and Excel. As a result, the organization didn't have a complete picture of all of the various contacts within prospect and customer accounts or the various interactions that were occurring with those accounts.

The marketing and sales teams worked well together to generate a large volume of leads at the top of the funnel. However, the sales team didn't have a good mechanism to prioritize their day-to-day work. Consequently, both lead and customer follow-ups often slipped through the cracks resulting in missed revenue growth opportunities.

Managers also expressed challenges in tracking individual rep performance. Variations in how each rep approached the sales cycle and differences in the definitions of where a prospect or customer was within the selling cycle made it impossible to forecast or to improve processes. Lastly, the team lacked consensus on how to measure the probability to close, which was further impacting executive confidence in the pipeline reporting numbers.

Global Beverage Solutions needed a common platform, standardized nomenclature, and a well-defined process that the team could agree upon in order to provide accurate pipeline forecast reporting for executive leadership and to continuously refine the selling processes.

Toolbox

The first thing to understand about mapping a sales process is that you might have more than one; in fact, it would be surprising if you didn't. This is largely due to differences in product lines or services, industries, external versus internal sales teams, and cultural differences if you're a global organization. It is important to define how many different processes you will ideally have moving forward. You should aim to have as few as possible, and they should only split paths where they absolutely need to be different.

Current State

If you are like many organizations at the start of the sales process design journey, you have as many current state processes as you have sales reps! While it may feel that way, you will likely find that they are all doing very similar things, perhaps just not using the same tools or in the same order or to the same degree as their peers. So, let's discuss how we go about defining your current state.

If you have eight reps or less, gather the entire sales team. If you have more than eight sales reps, gather a sampling of the various personas on the team. Plan to use large post-it notes and sharpies to collect information from them. You'll start by asking them to jot down what they are doing at the beginning, middle, and end of their current state sales process. They should also document any exceptions to the process. Since they will each be documenting their individual processes, they should use a post it for each major step.

Once they have them all captured, ask them to hang them on the wall in order of execution beneath pre-hung headers labeled Beginning, Middle, End, and Exceptions:

Beginning	Middle	End
Gather data about lead	Provide quote	Order confirmation
Identify influencers	Product sampling	Post close follow ups
Determine budget	Conduct product trial	
Formal presentation of company and products	Post trial follow ups	

At this point, the team can begin to compare and contrast where some reps may be completing more or less tasks than others or where they may be completed in a different order. Give the team time to discuss why they may or may not be doing some tasks, or why they're completing them in the order they are. If the reasons relate to differences in product lines, industries, or cultural differences, this is likely ok. You're trying to find both the existing commonalities and where there should be variances.

Pain Point Identification and Ranking

Next, you want to focus on the pieces of the current state that feel inefficient, broken, burdensome, incomplete, or painful. Give each teammate some red stickers and ask them to place them on the current state post-its that are particularly challenging or inefficient. They can mark as many items as they want to, but they must be marking something specific. Do not allow them to just say the entire process is painful. If you do that, it will be tough to decide upon concrete areas for improvement.

Be sure to walk through each item that was designated as painful. Capture detailed notes from the perspective of the person who tagged the item but also from their peers who may have a different perspective. You may find that the reason a particular task feels painful is because the person just needs more training, or you may find that several team members are struggling, and the item really does warrant further investigation. Once you have determined which items are truly pain points that need to be resolved, work as a team to prioritize the items in order to identify which items should be solved first, second, third, etc.

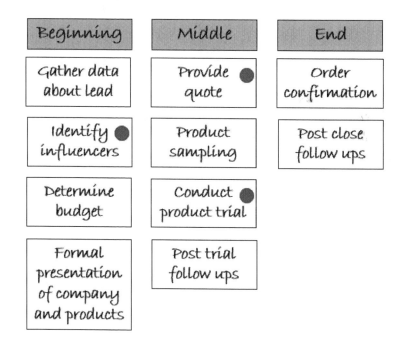

Pain Points		Vote
Identify influencers	2	●●
Provide quote	5	●●●●●
Conduct product trial	3	●●●

Root Cause Analysis

If you want to take the pain point discussion a level deeper to get to the heart of what is causing the pain, there are several techniques that can be deployed at this point. Note that in each of these techniques, there can be more than one cause to the painful effect you're feeling. Do the best you can to get to the root cause of each one.

Fishbone Diagram: Used to establish cause and effect. The effect is the current pain the team is feeling. The effect is placed in the center of the diagram, and the team calls out the various causes from there. You can add as many bones to the diagram as necessary to reach one or more initiating causes.

Use a template similar to this one to flesh out your fishbone:

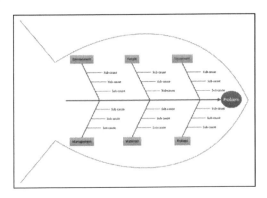

Pareto Analysis: Also known as the 80/20 rule, the Pareto Analysis states that 20 percent of causes contribute to 80 percent of your issues. In this method, you list your causes across the bottom, with the more urgent causes on the left. Urgency is determined by how many times you see the same cause contributing to multiple issues. The goal is to identify the Vital Few items you need to solve as compared to the Trivial Many.

The image below demonstrates a Pareto Analysis for a company that was experiencing errors with its website:

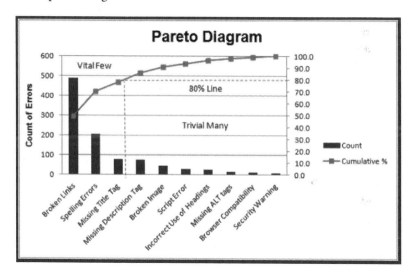

The 5 Whys: This method is popular because it is simple to conduct in the moment with very little preparation. Once you have the problem identified, you ask "Why" five times. This example outlines the cause and effect of a customer service issue for a printing company.

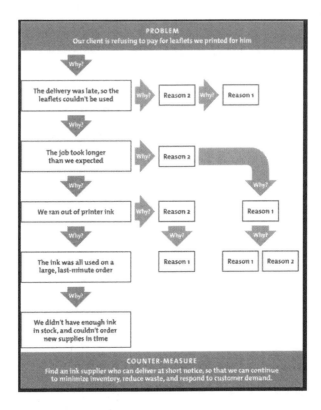

Pain Point Resolution

Now that the team has identified what is painful and perhaps the root cause of why you're experiencing pain, it's time to brainstorm some possible solutions.

List each pain point/root cause on the wall for the team to see. Ask each team member to individually brainstorm some ideas and post them on the wall. Every single idea the team generates should be posted under its parent pain point and discussed. You never know when one shared idea will spark a thought by another team member that could ultimately solve the problem. Make sure to weigh the pros and cons of each idea. Will it solve the issue in its entirety or only a portion of the pain point? Is it feasible from a timeline and budgetary standpoint?

After a thorough discussion of each item, work as a team to prioritize the solutions. You are aiming for which solutions should be implemented first, second, third, etc., considering the cost versus benefit of each one. Ideally, you will have some fairly simple quick wins that you can knock out early, along with some larger projects to complete over a period of time.

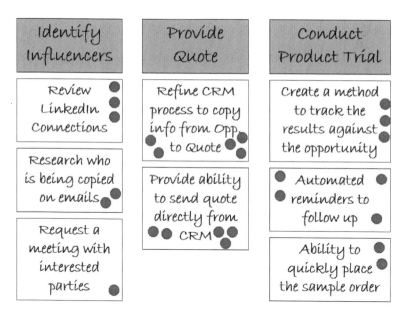

Envisioning the Future State

You've done the hard work of examining where you are currently and determining how to solve some of the pain points in the process. Now, you're ready to begin mapping out the future state sales process. Don't worry about the technology (i.e., "how" you're going to complete the steps). Just focus on what the steps need to be.

As discussed at the beginning of the chapter, you should aim to have as few processes as possible, and they should only split paths where they absolutely need to be different. So, the first step you need to do is decide on the minimum viable number of processes you need to define. Group the sales reps together that will be best suited to define each of the various future state processes you will need, but keep all of the groups in the same working space, because you want the processes to overlap as much as possible.

Start documenting the new process from the beginning, and work your way through to the end by moving items from your current state process over into the future state in the appropriate order. Gain consensus on any improvements you are going to be making that were gathered as a part of the customer journey mapping exercise and your current state exercises. In some cases, the improvement may replace an existing step, or it may include a value-add item that doesn't exist today. In either case, place those items in the appropriate order. Be certain each team is using consistent nomenclature and has the same definition for each step throughout the processes.

Once all the steps have been defined, work together to determine the natural stage breaks. A stage should have very clear entrance and exit criteria and not be dependent upon another stage. Once those have been delineated, come up with stage names that are meaningful for all the teams across the shared steps. Then add any additional stage names where the process may follow a different path, whether it be temporary or for the remainder of that particular process. Finally, mark any items that are required for the rep to complete before moving on to the next stage in the process.

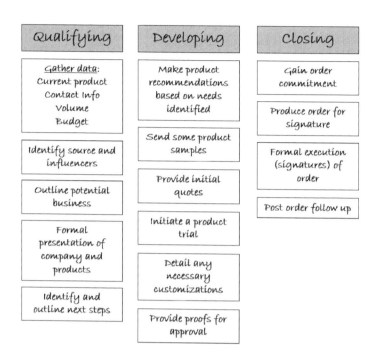

LUCK Alignment

If you want to take your future state process to the next level, ensuring you're developing a comprehensive process definition work through aligning each stage and its steps into the LUCK Process Matrix. The matrix pushes you to think about all aspects of a business process: data, decisions, deliverables, and ongoing development—The Four Ds. To do this, you simply need to add some row labels to the wall with the 4 Ds on them. Move your stage headers over as headers in your new matrix on the wall.

Next, you'll move each step in that stage into the new matrix, but before you do, number the steps within each stage in your existing matrix (i.e., Stage 1, Step 3). You'll be placing each item into the new grid where it aligns with both the proper stage and one of the 4 Ds.

As you look at each step, ask: is this a piece of incoming data, is this a decision, is this an outgoing deliverable, or is this a continuous improvement development item? Ideally, you keep the steps in order as best you can.

Once you have moved all of the existing items, review the new matrix for gaps. The most mature process is going to have a least one item in each stage/4-D intersection of the matrix. If there are gaps, work together as a team to determine what might be missing to fully flesh out the future state. Determine where in the order of the existing steps any new items should fall and whether or not those items should be required.

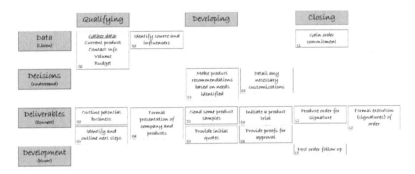

Successful Application

We last left our client, Global Beverage Solutions (GBS), with a challenge on its hands. The client's team needed some assistance in selecting and implementing a common platform, standardizing nomenclature, and defining a sales process that the team agreed upon in order to provide accurate pipeline and forecast reporting for executive leadership. As a result, GBS engaged an external partner in a two-prong approach, beginning with a Roadmap Planning Assessment, followed with a Sales Focused Phase One CRM Implementation project.

The roadmap assessment allowed the team to focus on outlining its current state processes, identifying the pain points, and beginning to envision an improved future state. The initial implementation focused on getting the sales team up and running using basic CRM accounts, contacts, and activities. This allowed them to add a much-needed structure to their day-to-day operations.

When it came to defining the sales process and opportunity management, the team was initially forced down the path of a seven-stage process by a member of the leadership team against the counsel of its partner. Once the process was up and running within the CRM solution, the team began to push back due to the time-consuming and complex process; as a result, the initial successful user adoption began to dwindle. The team was overwhelmed with the number of stages each opportunity had to move through and the sheer number of steps it had to execute within each stage. Many steps were required, and reps complained that they could not always get the required information in order to advance the deal through the pipeline. The process was too rigid and had put the team in handcuffs to a certain degree.

After a few months of declining user engagement, GBS re-engaged with its partner to reassess the sales opportunity management process. Together, they redefined a process that was only three stages, with an optional fourth escalation branch if a rep wanted to offer specialized pricing. The collective team was able to reduce the number of steps in each stage significantly as well as reduce the number that were required to move between stages. As a result, the process was now flexible enough to allow the team to collect the information in a more natural way based on how the relationship with the prospect was progressing.

Six months after the realignment, users became reengaged. The simplified process reduced the friction of tasks by ensuring everyone always knew what to do and when to do it. Managers could easily see which reps were working on each opportunity and more accurately report on the opportunity funnel and forecast.

CHAPTER 3
Connect

Technology Blue Printing

> "Instead of using technology to automate processes, think about using technology to enhance human interaction."
> — Tony Zambito

> "Just because digital technology makes connecting possible doesn't mean you're actually reaching people."
> — Maureen Dowd

The Challenge
Client Information:

- Commercial and Personal Insurance Managed Services
- >1,000 Employees
- Existing CRM solution in place

Managed Insurance (MI) has been in business for nearly seven decades. It is accustomed to delivering exceptional client service via its proprietary delivery model. The model is considered best in class in the industry. It is a robust process that allows account managers to engage clients with everything from uncovering their needs to working together to continuously improve their program over the life of the relationship.

The organization implemented a CRM tool three years ago but was struggling to fully maximize usage of the tool, so it took some steps to begin adding pieces of its proprietary model to the platform. However, it quickly found that this didn't improve the usage of the tool or provide any real benefit to the process. Account managers continued to use offline methods to capture client information and interactions.

The account managers were resisting change, because the bulk of the work they needed to execute within their process wasn't housed within CRM. They were still having to work within several different platforms and use Excel templates to capture the bulk of the information necessary to formulate the right program.

The team was also struggling to determine which of its client interactions should actually be captured within CRM and which were trivial enough to be managed elsewhere.

Another challenge the team was facing was that the actual service delivery team worked primarily on another platform. Each team had expectations of the other, but they didn't have a good mechanism for communicating those needs and sharing additional information.

MI was sure that it had the correct delivery model (what they called "the customer journey"), because it had been realizing success for years engaging clients with the model. However, the business process behind the model was inefficient, required too much manual communication, and frequently required rework. MI knew they needed a more automated process but didn't know if their CRM solution was the answer.

Toolbox
Technology Selection

Implementing technology to support a business process can be a daunting task. If you don't choose the right platform the first time, it can be a very costly mistake. There are several costs to be aware of when getting started on the technology path. The first are associated with selection costs as you begin to evaluate and ultimately choose the platform. Next to consider are the actual implementation costs, or re-implementation costs, if you didn't choose wisely the first time. Once the project is underway, you'll be exposed to customization costs, unknown requirements costs (aka change requests), potential third-party product expenses, and sometimes the cost of having two systems up and running in parallel for a period of time.

And then there's the human capital investment necessary to support the project during requirements gathering and testing. Once you're finally ready to release functionality, you're facing training or retraining expenses, followed by the cost of the disruption to daily routines within the business until the new platform becomes institutionalized. Finally, from a long-term perspective, you'll have costs associated with ongoing support and periodic upgrades. All that to say, conduct the due diligence to ensure you're making the right choice the first time.

Solution Demos

There are a couple of useful tools to aid in the selection process. The first is the Solution Demo which can be helpful in making the connection between a capabilities conversation and what those capabilities really look like. However, there are some considerations that need to be made when venturing down the demo path.

Know WHY you want or need a demo
Putting the next shiny object in front of your team to distract them from the current less than optimal reality is not a viable reason to request a demo! However, if you're brand new to a certain type of technology (read that as your team has never been exposed to it before), then a demo could be a great way to determine if the solution could possibly be a match for the business problem you're trying to solve.

Another perspective on this is to understand why *not* to request a demo. Most businesses approach a demo as the key step to take before selecting a solution. This approach puts the proverbial cart before the horse and frequently results in a failed project. See the section below on WHEN to ask for a demo for more information.

Know WHAT to ask for
Many platforms in the marketplace today offer multiple modules, basic and advanced features, and out of the box versus add-on third-party components. Be certain to clearly define what the business problem is that you're trying to solve for your existing or potential partner. Otherwise, your team could be inundated with information that may not be applicable to your needs, causing confusion when you're trying to compare and contrast solutions later.

When it comes to searching for the right CRM solution for pipeline and opportunity tracking, not only are there the heavy hitters like Microsoft and Salesforce, but there are also a slew of other providers out there that are considered add-ons to ERP platforms or standalone industry-specific options. Then within those solutions, there are modules like Sales, Service, and Marketing. And diving further, there are features offered out of the box within those modules, and there are third party add-ons for Sales Cadence Automation, Marketing Automation, Activity Management, and Document Management to name just a few. The possibilities can be quite overwhelming if you're just starting out, so be sure to ask to see only what your partner thinks you'll need to solve for the immediate needs. Once you've made that determination, you can explore further.

Know WHEN to ask for the demo

When a demo is provided can be just as important as why you're asking for it. As we mentioned before, demoing shouldn't just be used to make a team forget about a less desirable current situation. A demo should be requested once you have a clearly defined future state sales process that you're trying to maximize via technology. If you don't know what the process is, you'll never know if the tool will be capable of supporting it.

Know WHO should be involved

There are two important aspects to consider when it comes to whom should be involved in the demo process. The first is who should be delivering the demo. Ultimately, you want a partner that has a great deal of experience with the solution. You also want to ensure the team member they've asked to deliver it isn't too technical and should be able to demonstrate the features that are of the greatest interest to your team in a manner that will resonate best with that team and the way they would be using the tool.

The second aspect to consider is who from your organization should be involved. Many organizations make the mistake of inviting only leadership-level folks because they are typically the ones engaged from a sponsorship, budgetary, or project leadership perspective. However, it is imperative to include the sales reps and any administrative support folks who will ultimately be using the solution. They are going to be looking for specific aspects of functionality and ease of use that leadership folks may not consider.

Know HOW it should be delivered

When it's possible to have the entire team together with the partner in the same room, that's great! But that is not necessary to have a successful demo. The meeting can be held using virtual meeting platforms like TEAMs, GoToMeeting, Zoom, etc. What makes a demo successful is the level of engagement! The demo should first capture your attendees' attention by giving them a brief overview of the solution, focusing on ease of navigation and general organization of the content. Next, it should provide the team with a glimpse of a "day in the life" of the various users that may be interacting with it to solve the business problem at hand.

Referring back to our organization looking for a CRM solution, the sales portion of the demo may first focus on an outside salesperson, then on a regional manager, and finally on the role of sales director.

Lastly, a demo should always have adequate time for questions throughout. The team may have specific questions about how a feature or component will address specific challenges present in its unique processes, and it is inevitable that the attendees are going to see things on the screen that the partner may not be planning to talk about. They should be prepared to address the question without too much detail if it's an aspect that doesn't pertain to the immediate needs.

Fit/Gap Analysis

The second tool to consider is a Fit/Gap Analysis. The purpose of a fit/gap analysis is to evaluate the tightness of fit of a technology application to a specific business need. This applies to: (1) CRM solutions in general (for example, which CRM solution is the best fit for you), (2) sub-segments of CRM (for example, how well are the sales, marketing, and customer service functional areas of CRM being leveraged today), and (3) potential new functionality for CRM (for example, should CRM be extended [xRM] to cover other functional areas such as expense reporting).

The fit is how well CRM will match the needs, and the gap represents the needs that can't be met by CRM.

Here's a quick illustration:

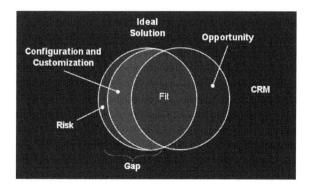

- You begin with a list of your ideal business process requirements
- Then, you assess the capabilities of the selected platform(s)
- Where you have overlap, you can see the basic fit
- Where there isn't overlap, you can see the gap(s)
- Some gaps can be reasonably covered with customization or a bit of custom development
- The risk is where you are unable to meet the requirements, even with custom development (or where the necessary development may be cost prohibitive)
- However, there's also opportunity where the solution may offer some unexpected benefits

The formula is different for everyone, because it is based upon your unique sales process needs. Practically speaking, what does that look like? Here is a fit/gap analysis for a loyalty system that had already been developed and was in use. However, it was causing redundant data entry for sales, because the team had to enter customer information and loyalty metrics into it annually, which had actually already been entered into CRM via accounts, contacts, activities, and opportunities.

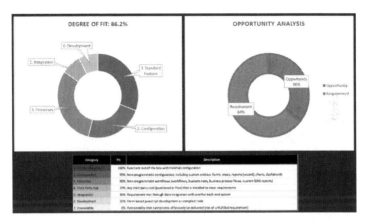

The Fit/Gap analysis showed that CRM could readily fit most of the requirements, with over 70 percent of the process needs being fulfilled without custom development. And there weren't any gaps in functionality that couldn't be covered at all, which meant little to no risk. Plus, there would be a 36 percent new functionality opportunity that the team wasn't asking for with mobile access, the ability to update data monthly instead of quarterly, and linkage directly to customer and opportunity information.

Technology Blueprinting

Once you've selected the platform and/or third-party product(s) that best fit your future state sales process, you need to blueprint the process to the technology. Think of blueprinting as part two of the customer journey mapping process. The blueprint is comprised of four major components and three types of interactions.

Data: Actions/Information that you're gathering from the prospect or customer

Decisions: Any necessary decisions either the customer or the business need to make to move the process forward

Deliverables: Tangible items that must be produced throughout the process

Development: Measurements taken along the way to manage the efficiency and effectiveness of the process

The customer engagement line shows the direct communications and actions between the customer and the organization. The process visibility line separates the front office activities that are visible to the customer from the back office actions that are not. The employee engagement line illustrates the separation of employees who directly interact with the customer from those that don't.

Develop your blueprint by layering your LUCK defined future state process into the various components of the template. (Download at: https://gotluck.link/Revenue-Engineering)

REVENUE ENGINEERING

- Begin with the customer journey stages
- Followed by your customer facing actions and supporting back office activities
- Next, list the deliverables
- Align the supporting technology with the actions and deliverables; be certain to consider the user interface of the devices that will be used by both the customer and your teams
- Denote any relationships between actions or tasks that may be deemed Stage Gate criteria
- Depict any integration relationships between technology solutions
- Layer on any expected timing guidelines
- Lastly, outline any additional metrics, such as probability to close

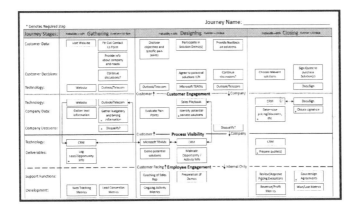

Successful Application

We last left our client, Managed Insurance (MI), with a challenge on its hands. The team needed some assistance mapping its proprietary customer journey model to the right aspects of its chosen technology stack. As a result, MI engaged a partner that understood both their CRM technology and sales process journey mapping. Together, they developed a three-step approach. They began with a Current State Assessment. This was followed by an enhancement project that resolved some foundational issues with the CRM implementation and then fully leveraged the solution to automate and streamline MI's successful model and business process.

The current state assessment allowed the team to focus on outlining its current state processes, identify the pain points within the technology, and brainstorm ideas to resolve the challenges. By outlining all of the various tools in play, the team could identify areas where niche applications could be replaced and where potential integrations between cornerstone platforms would improve the flow of information, reducing friction in the process.

The enhancement project focused on reengineering CRM leads, accounts, and opportunities. Mapping of fields across entities reduced duplicate data entry, while also keeping relevant information front and center throughout the entire sales process. Careful application of stage gates, required fields, and service level agreements provided structure and guidance to the business development team and an opportunity for leadership to coach and continuously measure the performance of the process. Special care was taken to optimize form design for easy data entry, regardless the device in play.

The simplified process, properly aligned with the right aspects of technology, enabled the team to be more efficient with initial data entry, ongoing interaction management, and sharing information within and across teams.

CHAPTER 4
Know

Peer Accountability and Process Improvement

> "Not everything that can be counted counts, and not everything that counts can be counted."
> — Albert Einstein

> "The wuss factor happens when a team member or leader constantly balks when it's time to call someone out on their behavior or performance."
> — Patrick Lencioni

The Challenge
Client Information:

- Health and Wellness Manufacturer
- >2,500 Employees
- No CRM solution in place

Health and Wellness (H&W) has been in business for nearly five decades. Its sales model includes a distribution network as well as some direct-to-consumer sales through outside sales reps. The direct-to-consumer sales team had been tracking its sales orders and commitments in Microsoft Excel, which was tough to share and report on. Leadership determined that the direct-to-consumer channel was a strategically important area to improve in order to expand their footprint and meet growth goals. For this reason, the business decided to launch a CRM project.

The hope was that a CRM solution would allow for easier tracking of prospects and customers. Team members were entering commitments into one system, orders into another, and re-keying some data into both—streamlining this into a single system that worked on either a desktop or mobile device was also an important goal.

The sales team had a go-to market framework it had been using thus far but had not mapped that process to the capabilities of a CRM solution. The critical component to the strategy revolved around face-to-face meetings with prospects and customers in order to demo the product. With the assistance of an implementation partner, H&W spent some time doing some upfront planning and then moved on to the actual implementation project.

The implementation consisted of basic CRM accounts, contacts, activities, and orders. To enable the team to efficiently capture orders into CRM, integration with H&W's ERP solution was also a critical requirement. The team got to work using CRM and was experiencing some success, since the team could now quickly and easily enter orders for processing from mobile devices. Reps were also logging post visit notes from their face-to-face meetings.

It was clear, however, that initial adoption of the platform wasn't meeting expectations. The team still wasn't using the platform to proactively manage customer relationships or track commitments. As a result, the leadership team couldn't easily identify where the reps may need some coaching or an assist in an upcoming meeting in hopes of gaining a commitment. H&W was unsure if the process was the issue, if CRM was missing a component the team needed, or if it was purely a training or adoption challenge.

Toolbox

A critical component of business process maturity is the ability to measure how well the process is performing. Numerous factors can be considered, but the most common include the following items:

Process Effectiveness measures the process deliverables against customer expectations.

Process Efficiency is the measure of how long it takes for the employee to conduct the tasks within the process.

Resource Productivity is the measurement of how effectively the resource is using their time.

Cycle Time is the duration of time the process takes from initial input to expected output.

When bouncing these ideas against the concept of a sales process, it might look something like this. Effectiveness relates to how well you are delighting the customer throughout the buying journey. Efficiency can be measured at any single point in the sales process—at the stage level or even at individual step level. Resource productivity can be helpful in determining whether the sales rep is focusing on the highest priority items and maximizing activity planning. Lastly, cycle time can be helpful to understand process and/or stage duration and probability to close.

One thing you can be sure of is that your team probably isn't going to like the concept of measurement, because measurement typically means accountability! And the fear is understandable. Without the right support and communication, accountability can feel like a threat.

"If the project I managed fails, I'll get demoted."

"If I lose an opportunity, my commission will be impacted."

"If I don't meet my quota, I'll lose my job."

Accountability is important if you want the individuals and the team to meet their goals. But making your department a safe place for accountability is an important part of the formula for success.

The last thing to remember, before we dive into the how-to of measurement, is that a house isn't built in a single day. You're looking to introduce small, incremental changes over a period of time that have a lasting impact on long-term behavior. Those changes in behavior should have a positive impact on your goals. However, we need to be careful about just how many goals we're chasing. According to Franklin Covey, when a team sets out to achieve 2–3 goals, it can reasonably achieve those goals. When a team increases its initiative to 4–10 goals, it can expect to reasonably achieve only 1 or 2 of the goals. In other words, the more goals a team sets, the fewer it actually accomplishes! So be mindful of that, set your sights on changing one or two behaviors at a time in pursuit of 2–3 annual goals.

How do you get started with effectively measuring the team, the process, accountability, behaviors, and goals? Much like the other concepts we've covered, you'll start with a workshop, so gather your sales managers and reps. An in-person meeting is best, but a web meeting can be just as effective, if need be.

Define the Measurable Objectives

You're going to start super simple and very high level. The first thing you want to do is determine what the measurable objectives are for the business and the team. Some things that might get mentioned here include increasing revenue, profit, and market share or lowering costs. The challenge with measuring these items is that they're results oriented, which means you have no idea if you've achieved the goal until it's too late to do anything about it! Results are known as *lagging* indicators, because the measurement lags behind the behavior.

Brainstorm the Behaviors

To get past the lagging indicator issue, work with the team to brainstorm the *behaviors* that could have an impact on each of the objectives you outlined. Let's use the increase revenue objective as our example. Some behaviors that might contribute to increasing revenue—which in reality means winning more deals—include visits with the prospect or customer, delivering demos, following up with phone calls or emails, sharing relevant product information at the right time ... the list could go on and on, depending on your industry, product, or service.

Complete this exercise for each of the measurable objectives you brainstormed in the first exercise.

Prioritize

In order to narrow your scope for the next step, work with the team to first prioritize the measurable objectives you outlined in the first step. Once you've chosen the most important business objective, prioritize the behaviors beneath that objective. You want to select one or two—remember, no more than two—behaviors to focus on.

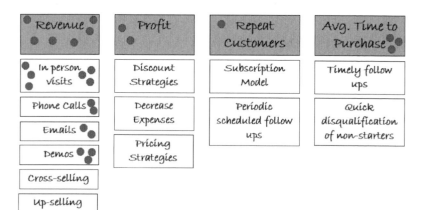

Determine the Metrics

Now, it's time to turn those behaviors into metrics, also known as *leading* indicators. Continuing with our example, if we turned our behaviors into metrics, they might look something like this:

- Increase in-person customer visits from [current performance] to [performance goal] by [when]
- Increase the number of demos delivered from [current performance] to [performance goal] by [when]
- Increase the number of follow up phone calls from [current performance] to [performance goal] by [when]

You may have noticed that everything you've done so far has been based on human behavior. So how do you begin to measure the actual sales process? Start by setting two baseline numbers for each stage of your process.

Probability: This is the percent likelihood that the deal will be won, given the stage it is currently in and the information collected thus far in the process. The earlier in the process, the lower the likelihood to close the deal; the later the stage and level of information collected, the higher the probability to close. You should not let reps determine this percentage on their own on each deal, because you are always going to have some reps that are very optimistic and some that are very pessimistic. It's best to let CRM set the probability, since it will have an impact on the weighting of your pipeline reports.

Duration: This is the length of time you believe each stage will take to complete. Depending on your product or service, this could be a matter of days, weeks, or months for each stage.

Over time, as the team works within CRM to track opportunities, a large amount of data can be collected with regard to these two measurements. You can begin to evaluate the actual probability and duration over the period of time and adjust the values in CRM to match reality, resulting in an even more accurate pipeline report. This has the added benefit of significantly improving the accuracy of your sales forecasts and cash flow forecasts.

Set the Goals

This is the hardest part! Let me introduce you to the concept of SQAGs™—Small Quickly Attainable Goals. SQAGs should be leading, not lagging. They should be a hundred percent within the control of the individual to achieve or not. They should be super easy to understand and able to be measured on a weekly basis. Lastly, the SQAGs should be sustainable and team based. While you are going to be holding individuals accountable for their portion of achieving the goal, you're going to focus on measuring the team's performance.

You should begin by determining the current baseline for each metric that you want to measure. Is the team currently making ten calls per week or a hundred? This will set the [current performance] variable in the metric statement we demonstrated earlier. Next, you'll work with the team to determine what the [performance goal] variable should be at three different levels—Quick, Target, and Long.

The Quick goal should be a small increase from the current baseline—a goal the team will be able to reach in a fairly quick period of time so that you can increase morale and garner support for the hard work to come.

The Long goal should be something that feels much like a performance plan stretch goal. It is really going to take some effort to achieve it, and you don't expect the team to be able to sustain it for a long period of time.

The Target goal is somewhere between the Quick and Long goals. It is the ideal place you would like the team to settle back to after reaching the Long goal. The behavior shouldn't fall all the way back to baseline or even the Quick level; the Target goal should be considered the new normal once the behavior has been institutionalized.

See the example below which represents: "Increase the number of daily in person visits by rep from 4 to [Quick, Target and Long goals] by 90 days from now":

Goal	Metric
Current Baseline	4
Quick Goal	6
Target Goal	8
Long Goal	10

Plan the Celebrations

Now that you've gotten through the hardest part, let's move on to the fun part! Work with the team to brainstorm a theme for your program. Then, within that theme, you're going to brainstorm how you'll celebrate the achievement of the Quick, Target, and Long goals. Be sure to set a budget for each level, and then focus on having fun together. While gift cards are nice, this is about building team dynamics and celebrating some tough work alongside one another. Think something like Deals and Diamonds as the theme with celebrations that include ballpark burgers and dogs grilled by the leadership team as the Quick celebration, tickets to the AAA team game as the Target celebration, and dinner and tickets to a major league game as the Long celebration. They don't necessarily have to be big on spend, but they should be big on fun!

LUCK*score*™—Roll Out the Program

You're in the homestretch! The last thing to do is roll out your LUCK*score*™ program.

Use the outline and templates below to get it going. (Download at: https://gotluck.link/Revenue-Engineering)

- Choose your start date and give yourselves 3–6 months to make it through all three goal levels, depending how lofty they are.
- If you're able to track and pull the information directly in CRM, this is the best bet. Just make sure you've got a dashboard set up to tally the results for you, both by rep and at the team level, on a weekly basis. If you aren't able to measure the data within CRM, create an Excel file for tracking and reporting. The reporting piece can be easily achieved using a Pivot Table, provided you include fields for Team Name, Rep Name, and Week Number. If you're going to be tackling more than one goal at a time, you might want to consider either two tracking sheets or an Activity Type field in your file.
- Prepare your Team Score Card, where you will track your metrics

				Week Ending				
				Week #1	Week #2	Week #3	Week #4	Week #5
OWNER	METRIC	GOAL	AVERAGE	4/3/2021	4/10/2021	4/17/2021	4/24/2021	5/1/2021
MONTHLY								
xxx	Revenue (MONTHLY)	$ -	#DIV/0!					
xxx	Profit / Net Income (MONTHLY)	$ -	#DIV/0!					
xxx	Gross Profit (MONTHLY)		#DIV/0!					
WEEKLY								
xxx	# Active Clients		#DIV/0!					
xxx	# New Clients		0 #DIV/0!					
xxx	Total # Opps, Next 6 Months		0 #DIV/0!					
xxx	Weighted Pipeline, Next 6 Months	$ -	#DIV/0!					
xxx	Estimated Pipeline, Next 6 Months	$ -	#DIV/0!					
DEFINITIONS								
	# Active Clients	xxx						
	# New Clients	xxx						
	Total # Opps, Next 6 Months	xxx						
	Weighted Pipeline, Next 6 Months	xxx						
	Estimated Pipeline, Next 6 Months	xxx						

- Schedule a weekly recurring meeting for 15–30 minutes or include LUCK*score*™ as an item in your regularly scheduled team meeting.
- Prior to the meeting, the metric owner(s) should ensure the score card has been updated with the most recent results.
- In the weekly meeting, the team should review its progress by asking the following questions.
    - How did the *team* do last week?
    - What was the trend?
    - Did *we* make a quick, target, or long goal?
        - If so, schedule the celebration date!
        - Celebrations should happen within two weeks of meeting the goal
    - What will *you* commit to doing in the coming week to improve the trend?
        - Each teammate should be prepared to verbally make a commitment to their teammates
            - For example, the team goal for the week may be 30 follow-up phone calls per day or 150 per week, and there may be six reps on the team. If you're going to divvy it evenly, each person needs to be making 5 calls per day or 25 per week. However, Lucy may be attending a trade event this week, so she may only be able to commit to making calls on three of the five days. The other teammates can either pick up the slack or ask Lucy to increase her call volume on the days that she can be active. Over time, the team dynamics will sort themselves out.
- Be certain to establish a culture of peer accountability. The team needs to know that it is absolutely within bounds to respectfully push back and challenge one another when it

comes to meeting the goals each person committed to achieving. It is also an excellent opportunity to share success stories and engage in some peer mentoring if the opportunity presents itself.

Once the team has worked through this 3–6-month goal, circle back to your list of brainstorm items to choose the next most important behavior to focus on. Ideally, you're working through one per quarter or two every six months.

When selecting the right metrics and proactively committing to their piece of the team goal, the LUCK*score*™ program also aids in shifting the mentality around sales process and CRM utilization from "Big Brother" to proactive planning tools that helps a rep do their job more effectively and efficiently. And the measurement of the process itself can help level set expectations around probability and duration so that reps know when to ask for help, and managers have better visibility into where additional coaching is necessary.

Successful Application

When we left Health and Wellness (H&W), it was feeling discouraged about the initial adoption of the platform. The team wasn't using CRM to proactively manage customer relationships or track commitments, and leadership had little visibility into where the reps may need skills coaching or assistance in upcoming meetings in hopes of gaining a commitment.

Since the H&W team was new to both a more defined sales process and to CRM, it chose a LUCK*score*™ program that would provide some insight into proactive behaviors and increase the adoption of CRM.

- The Program: Champagne Wishes and Caviar Dreams
- The SQAG: Increase preplanned face-to-face prospect/customer demo visits
- Current Baseline: 108 per week
- Quick Goal: Increase preplanned face-to-face demo visits from 60 to 100 per week by the end of the quarter
- Target Goal: Increase preplanned face-to-face demo visits from 100 to 150 per week by the end of the quarter
- Long Goal: Increase preplanned face-to-face demo visits from 150 to 200 per week by the end of the quarter
- Quick Celebration: Share a bottle of champagne and hors d'oeuvres
- Target Celebration: Team dinner at a nice restaurant and shared champagne
- Long Celebration: Team dinner with significant others at a nice restaurant and a bottle of champagne for each person to take home

Due to hectic travel schedules, the team was unable to meet weekly, so they met once per month and measured the goals on a rolling three-week schedule. They also combined their in-person meetings with a digital social collaboration platform (Microsoft Teams) to share ideas and hold each other accountable in off weeks. As a result, it took a little longer for the team to gain momentum and meet the Target and Long goals, but it eventually rallied and was able to make all three.

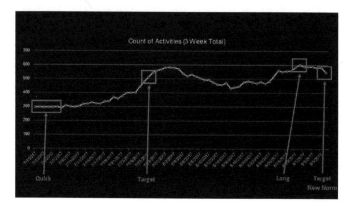

CHAPTER 5
Good LUCK

Culture and Change Management Includes Leadership Engagement

> "The way management treats associates is exactly how the associates will treat the customers."
> — Sam Walton

> "How do you know how to improve a business? Easy. Do the job on the front lines."
> — Marcus Lemonis

The Challenge
Client Information:

- Public Services
- >5,000 Employees
- Existing CRM solution in place

Service Central (SC) has been in business for more than four decades. It is both a business-to-business and business-to-consumer operation with both inside and outside sales teams, a customer service team, and very active marketing department. Since the organization is rooted in the public services industry, finding opportunities to sell its services wasn't the biggest challenge it was facing. The majority of the pain points related to business process and how to maximize its investment in CRM to prioritize work and streamline handoffs.

The initial investment in CRM yielded the typical results related to easier tracking of prospects and customers as well as the ability to create incoming leads and convert them to opportunities. However, due to a multitude of product lines and services and a complex pricing model, the team was struggling to generate quotes and contracts in a timely manner. The business was using a combination of CRM, Microsoft Excel, and Microsoft Word templates to execute most deals.

Due to complexities in its data model, the marketing team was also struggling to track a lead all the way through to a closed opportunity. As a result, it had no idea which campaigns were effective and which ones weren't. And due to the number of leads being received via phone and web inquiries, the teams sometimes didn't follow up quickly enough (or at all), resulting in losing business to a competitor. Once a customer had been won, the customer care team didn't have a good way to track incoming issues with service. Since most issues were the result of trouble with a product or service, the ability to coordinate with the operations group was paramount.

Given all of the challenges it was facing, SC knew it needed some assistance, but it wasn't sure exactly what that would mean for the processes it currently had in place or for the configuration of its existing CRM platform. SC engaged a partner to conduct an assessment of its current state as well as a technical health check on the CRM environment. The result of those engagements was a roadmap that helped SC evolve both their processes and technology forward. Without the active engagement of the leadership team, however, SC's efforts would have been unsuccessful.

Toolbox

When it comes to onboarding and institutionalizing business process, especially business process supported by any technology solution, leadership engagement is about so much more than sponsoring the project and signing the checks. Since this book is focused primarily on sales process design, we're going to speak about leadership from the CRM perspective.

Ideally, before the CRM initiative is even launched, your organization should have pulled together a steering committee. This committee would be responsible for a lot of different things over the course of the CRM and sales process lifecycle. Depending on the size of your organization, this committee might consist of a few key players, or it may consist of a central committee with several sub-committees.

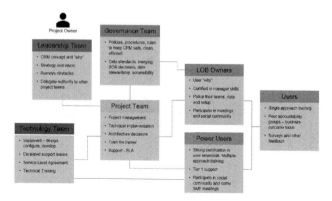

Communications

First and foremost, this committee should be on the hook for ensuring there is an adequate amount of communication occurring. A communications plan should be drafted to deal with any project work occurring—whether it be an initial implementation or how ongoing enhancement work will be shared. The plan should include an introduction to the work, notices regarding upcoming meetings, and/or progress reports along the way. The communications should ramp interest and excitement about the features that will be coming. Also, make sure the messaging includes the WIIFM hook—"What's in it for me?"—so that employees will understand how they are directly going to be impacted.

Finally, communications encompasses the feedback loop. Make sure employees understand how to provide feedback and that you have a way to follow up on that feedback in a timely manner. As you're building out the communications plan, make sure you're selecting the right channel for each type of communication and/or the audience the information is being shared with.

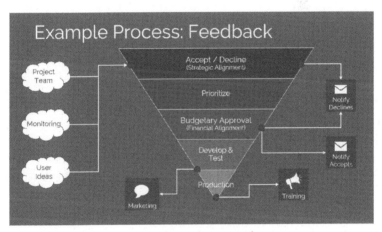

Change Management

Change management relates to project management, technical functionality releases, and human beings. It is important to have rigor around all of those aspects. However, for the sake of this chapter, we're going to focus on change's impact to your human capital. Most human beings resist change, which means you need to deal with change management head on.

There are several tried and true change management methodologies, but when it comes to dealing with people, Prosci's ADKAR model is spot on, because it gives leadership the tools it needs to facilitate change while supporting employees throughout the lifecycle of the change.

Awareness: Remember "what's in it for me?" that was discussed in the communications section above? This is a crucial aspect of creating awareness. However, the major distinction you want to ensure employees understand is the awareness of the *need* for change, not just awareness that a change is coming. To do this, be sure to focus on the reason for the change as well as the risks associated with not changing. Leadership should be actively engaged in communicating the awareness messages, as well as throughout the duration of the change acting as change agents, modeling the desired behavior. Leadership should also be engaging mid-level managers and team leaders to ensure employees are hearing consistent messages throughout the organization.

Desire: This is perhaps the greatest challenge that you will face when it comes to change management, because you cannot create the desire in someone to embrace change. You have the ability to be a change agent by having a positive impact on the impending change, but you cannot decide for your team members whether or not they have the desire to embrace change. Some factors that your employee is likely experiencing that you should be aware of so that you can consider how to support them include:

- A fear of whether the change represents an opportunity or a threat to them or their position.
- How they perceive the organization's ability to execute change successfully, based on previous change initiatives.
- Whether they are dealing with a significant amount of personal change, or if the change you are proposing is directly opposed to a personal belief.

- What their individual personal motivators are—some people thrive on recognition, while others may simply value relationships.

Knowledge: This milestone impacts an employee's knowledge of what to do during the actual change as well as what success looks like in the future state, post change completion. For example, do employees have the right information to manage through the various stages of the change? Will they be needed for any meetings? Will they be required to provide feedback? Will they have to manage their work with a foot in two different worlds for a period of time? Will their position or job responsibilities be different? Will their business process or tools be updated?

Ability: The biggest thing to remember about change is that while people may be aware that the change is coming, have the desire to do it, and have the knowledge of how it is supposed to be done, that *does not* mean they necessarily have the innate ability to execute. It will be very important to ensure that your team has access to mentors and subject matter experts. They will also need a safe place to practice honing their new skills.

Reinforcement: It will be important to hold the team accountable for adopting the change, praising individuals for work done well, and providing constructive criticism to help them grow. However, understand that not everyone is going to engage and adopt at the same pace, no matter what approach you take to rolling out the change. As discussed in the Know chapter, incorporating recognition, rewards, and celebrations when individuals or the team hit milestones will be important in building momentum.

For more information about ADKAR visit www.prosci.com/adkar.

Implementation Roll Out

Just as no two customers or even opportunities are the same, no two business process or technology implementation projects are the same. Thus, the characteristics of the solution, the business process, the team, the timeline, and whether or not it is an initial implementation or an enhancement should be weighed before selecting a rollout methodology. There are several popular approaches available, between LEAN, Six Sigma, Waterfall, Agile, etc.

When it comes to implementations or enhancements where user adoption is critical, a rollout weighted with Agile practices is best. Agile's greatest benefit is early engagement of critical employees and continuation of the feedback loop throughout the development cycle.

Agile projects consist of iterative releases called sprints. Each sprint engages the necessary stakeholders who work alongside the project team to ensure that the correct requirements are identified, conduct thorough testing, and report feedback. Each sprint consists of a looped process that entails prioritization, requirements gathering, delivery, and demonstration to senior stakeholders, feedback/rework, and approval of the deliverables.

The benefits of this process include engagement of the necessary employees early and often, the ability to tackle issues quickly as they are identified, and firm management of the project timeline and budget. The prioritization portion of the sprint loop aids in managing the budget and scope creep, while the demonstration enables project managers to hold the team accountable to deadlines. Additionally, frequent delivery of manageable portions of system functionality assists employees with the learning curve. Ultimately, this aids in mitigating one of the largest project obstacles—user adoption.

Trickle Training 🍀

A tried-and-true methodology called STARS, developed by C5 Insight, was designed to provide a multifaceted approach that ensures trainees have a successful training experience that leads directly to successful adoption. The best results are realized when STARS is trickled out with small, bite-sized chunks of knowledge. When training is compressed into 1–2 intense days (a.k.a. drinking from the firehose approach), it can be overwhelming, and knowledge retention can be low. For example, when rolling out and training on a new CRM implementation, a sales team may first focus on learning how to enter its accounts and contacts; a month later, it may focus on opportunity tracking; another month later, it may be trained on activity management.

STARS is comprised of several different teaching methods that work together to provide a comprehensive training methodology. It consists of five high-level steps, as shown in the diagram below.

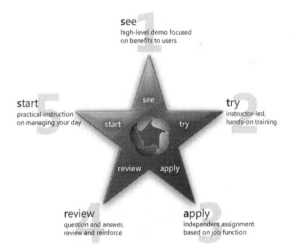

Note that STARS provides a proven framework for successful training—but the specific content of each step varies based on the requirements of each implementation. A description of each step is included below:

See: The keys to a successful "see" portion of training are to stay focused on the big picture without diving into detailed "how-to's" and to provide a highly relevant, but brief, demo targeted to a specific group of employees.

The See portion of the STARS approach is the lecture-based segment of the training. Although many employees are eager to jump directly into hands-on activities, it is critical to provide trainees with the big picture of how the application works before diving into the details. When focused on the benefits to the employee through a "day in the life" or scenario-based demo, this segment of the training is also useful for building excitement around the process or tool.

The demos should be brief—around 30 minutes—and should be focused on "selling" to the attendees by showing the key benefits. A good approach is to use a targeted-use case showing how a typical employee would successfully use the process or tool throughout their day.

Try: Once the trainees understand the high-level concepts of the process or system, the hands-on part of the training begins. This segment often contains the most substantial content of the training session. The instructor covers a step-by-step walk-through of the knowledge and skills needed. If possible, the trainees try it themselves by following along.

Apply: Busy salespeople, managers, and executives often leave training with the best of intentions but take weeks or months to apply what they have learned. Research from Work-Learning Research shows that people forget as much as 90 percent of what they have learned after one month. The Apply step helps trainees lock in their knowledge immediately following the initial training.

During this segment of the training, the employees are provided with specific assignments and are encouraged to complete the work individually. This portion of training may be completed in class, with instructors available for assisting, or it may be provided as "homework." Typically, this is the time most of the actual learning occurs because employees are going through real-life daily tasks and responsibilities. The content of the assignments should be as relevant as possible (it is preferable to use actual assigned tasks so that real work is being accomplished) to the employees and their day-to-day use of the process or system.

Review (and Recognition): Once the Apply segment has been completed, the training group is reassembled for a period of question and answer, review, and reinforcement of key concepts. The review should consist of reviewing the assignment itself, running adoption reports, providing recognition to those who excelled at the assignment, and reinforcing the learning by answering questions raised by employees. During this session, we also recommend giving some recognition to those who have done the best job of adopting the solution.

Start: The final step in the STARS training approach is the one that is most often overlooked. Employees typically leave training with adequate technical knowledge and skills but lack the practical understanding of how to perform newly refined job functions. The Start portion of the training focuses on providing concrete instructions for managing everyday work using the new process and tools. A good way to approach the start portion of the training is by prioritizing the daily activities of a typical employee. Then, by explicitly demonstrating how to perform each function, employees have a better understanding of how to get started.

Culture and Leadership Engagement

Corporate culture. We know that great ones produce great products, great places to work, and great customer experiences. But what causes cultural problems, and what are some practical steps to take to turn the corner? The theory behind cultural challenges is that an imbalance exists somewhere within the four pillars of the organization. Maybe there are issues with micromanagement, with trust, with communications, with training . . . the list goes on.

The four pillars of an organization:

Maturity: The length of time it has been in business, and how developed its processes and procedures are.

Mission/Vision: Why a company is in business, and the values by which it chooses to operate.

Culture: The day-to-day hum of an organization; you can feel it as you walk the halls or interact with individuals and teams.

Structure: A deep dive into the org chart and whether or not the firm chooses to interact in a hierarchical nature, a matrix, or a more flattened, collaborative format.

The solid or dotted lines in the org chart and between the other three pillars reflect the level of engagement within and across those pillars. When the energy or engagement of the company is not flowing well between those four pillars, the performance of the organization begins to suffer.

Symptoms of Low Leadership Engagement

There are many things that could cause an organization, or even just a subset of an organization, to suffer from low performance. Let's follow a few of the most common symptoms through a remediation plan:

Symptoms

- Decreased customer satisfaction
- Lower revenue
- Communication decreases
- Silos increase
- Negativity spreads
- Increased employee turnover
- Increased costs

Potential Causes

Symptom	Potential Cause(s)
Decreased customer satisfaction	Low or slow innovation practices, lack of connection to the buyer personas and their journey, inadequate processes in place to serve the customer, disengaged employees
Lower revenue	Decreased customer satisfaction, lack of focus on core competencies, poorly conceived sales, marketing, and operations strategies
Communication decreases	Subpar information sharing both vertically and horizontally in the organization, lack of a closed loop feedback process, lack of fresh perspectives
Silos increase	Too many stovepipe systems, tools or apps, habitual responses to repeat problems, lack of cross-team collaboration
Negativity spreads	Lack of accountability, resistance to change, fear of speaking up
Increased employee turnover	Ineffective change management practices, lack of communication, relying too much on internal resources as the only source of leadership
Increased costs	Inability to discern negative patterns and learn from them, this is how we've always done it, forgetting to ask why, lack of root cause analysis

Creating the Unified Ecosystem

So how do we get back to an organization with positive, proactive leadership engagement coursing through, to, and from the four pillars? Take a look at the checklist below to see if your firm, department, or team is falling short in any one of these areas.

Mission/Vision. Make sure you have a clearly stated mission and vision that has been communicated (frequently) throughout the entire organization. Not only should it be communicated, but leadership should take steps to ensure that each and every role in the company understands how it contributes to the bigger picture.

Every organization exists to serve a customer, so invest time in truly understanding who your customer is (sometimes, they're internal). Develop personas and map out their journey(s) all the way from prospect to post-sale support. Would *you* want to do business with you?

Listen to your employees. They need to feel like they're making a difference, and if they don't, they probably aren't. Be open to suggestions and be certain to provide feedback, even if it means telling them, "Thank you, but we can't tackle that right now."

Know your strengths AND your weaknesses. Revisit strengths and weaknesses every year and determine where it is feasible to take on new challenges. By the way, remember there shouldn't be more than two or three at any given time. You can learn a lot from failure, so be vulnerable enough to investigate what happened—and why—by conducting lessons learned postmortems after each and every project or initiative.

Capture information where it matters. Use the tools at play in the organization to collect, organize, and report on important information.

Do not ask employees to capture information in one tool just to export the data to another for consumption. Become a user of the tools your employees are using. Learn how to maximize the information being collected to drive your meetings, your decisions, your coaching opportunities, and your own customer interactions.

Break down barriers. Eliminate the walls between systems, tools, and apps by integrating or eliminating them where possible. Promote cross-team ownership of ongoing projects, processes, and technology solutions. Connect leadership and the front lines by using CRM tools and social features. By giving employees fewer places to go to share and find information, they can spend more time on things that are important instead of too much time searching for what they need to get their job done. The things they will learn from one another while collaborating could spark the next innovative idea for the organization.

Embrace change. How comfortable an organization is with embracing and working through change speaks directly to leadership and culture. Nothing in business is constant except the challenge of meeting a customer's needs. To that end, once you have defined/refined a process, don't just set it and forget it. You should always be looking for ways to improve, and that means you should always be measuring performance, and not just the performance of employees and things like revenue, expenses, and profit; it's too late to do anything once you have those numbers in hand. Focus on measuring the behaviors that drive the performance you're looking for throughout the entire organization. Set goals and hold one another accountable to meeting them.

Don't forget to celebrate. This is important. Your employees are human beings, and you need to treat them that way. That means stopping the daily grind to recognize the effort behind the achievements and celebrate at work. Do it as a team, because you're all in it together. This helps people connect to that sense of purpose we talked about earlier, and it further enhances the relationships amongst colleagues.

It's a lot, but you'll get there. Examine where you are, develop a roadmap of where you want to go, and outline the plan you'll take to get there. Make sure to refine it along the way, as the business evolves, and you learn from mistakes.

Successful Application

We last left our client, Service Central (SC), with a challenge on its hands. The team needed some assistance streamlining its business processes, reducing the friction in handoffs between teams, and maximizing its investment in the CRM platform. As a result, SC engaged with a CRM partner to examine how its current state processes fit against the CRM platform and to assess how it could better facilitate cross-team collaboration.

The process assessment allowed the team to focus on cleaning up the forms each team was using and the fields that were relevant to each persona on those forms. The team also focused on which views, charts, and dashboards were the most meaningful and removed the ones that weren't. All of those items aided in reducing the amount of clutter any particular persona was weeding through when they worked in CRM.

When it came to assessing better team collaboration, the team became laser-focused on the chain of events that need to occur from the time a lead is captured all the way through opportunity closing. In doing so, the inside and outside sales teams, marketing, and customer care became more aligned in delivering an excellent customer onboarding experience. Some adjustments had to be made to alter the original architecture to support better lead source tracking. From there, a business process flow was deployed that pre-created some sales cadencing activities for the inside sales team to manage, including follow-up activity tracking.

Should the need arise to hand a lead off to the outside sales team, the inside team had a quick efficient way to set an appointment for the outside sales rep without having to check with them first. Once the opportunity was underway, the sales teams had an easy three-stage sales process to work through that incorporated quotes and the usage of a third party configure-price-quote engine. This engine allowed them to produce quotes on the spot for interested prospects rather than having to reach out to another team to obtain pricing for services. If the customer approved the quote, the team could then easily create the service agreement proposal and send it using integrated e-signature capabilities. Because CRM had been integrated with SharePoint, the quote and proposal documents were easy for the customer care team to access without requiring access to CRM so customers could be onboarded seamlessly.

All of this was possible because the leaders of each team listened to their employees and worked alongside them to determine where the bottlenecks were occurring, and the leadership team was vulnerable enough to recognize that the organization was going to need some help from an outside vendor to tackle some of the challenges it was facing. Because the leaders were engaged and vulnerable, the employees trusted the decisions being made, engaged a hundred percent in the change management journey, and wholeheartedly adopted the updates in functionality. The team is now pursuing some advancements in campaign automation to proactively communicate with its customers and prospects more often.

CONCLUSION

Revisit Annually

So, you've done it—you've jumped in headfirst and worked through designing a sales process. But you're not done yet! You should plan to revisit the process at least annually. One of the greatest benefits of a sales process is that it is easier to evaluate and improve sales results once it is in place.

Start by reviewing the personas you developed in the Listen chapter to ensure they still align with how the market is conducting business. Review the customer journey as well to see if you have the ability to reduce any new pain points that may be popping up or introduce a new wow factor that you're now able to deliver.

Next, it's important to review the process metrics you established in the Know chapter to see how accurate your data is and how well the process is performing. As more and more data is captured, your metrics should become easier to use in forecasting. As for the process itself, you might find that you need to engage in some course correction along the way. If you're seeing poor performance, conduct some workshops to review your front-end processes from the Understand chapter and back-end systems from the Connect chapter to identify inefficiencies, pain points, and new opportunities to take advantage of.

Tackle Other Processes

Many steps in the book can be used to design processes for other customer-facing teams like customer care, technical services, field services, or project services. Simply work through each chapter from start to finish with the right hat applied!

LUCK Resources

We have all kinds of resources to help you put LUCK into practice with process design and other initiatives, and we're continuously expanding the collection of items.

If you would like to bring LUCK to your company or event, the team at C5 Insight provides advisory services, planning, training workshops, talks, audits, analysis, implementation, bulk book specials, and inspiring keynote speeches for companies and organizations around the globe.

We can help you become Powered by LUCK! Visit www.C5Insight.com for more details.

REFERENCES

HubSpot
https://blog.hubspot.com/sales/sales-statistics

Nielson Norman Group
https://www.nngroup.com/articles/service-blueprints-definition/

Prosci founder Jeff Hiatt
https://www.prosci.com/adkar/adkar-model>

Qvidian
http://thumbnails-visually.netdna-ssl.com/the-top-challenges-facing-sales-leaders-in-2014_52ea76b2229a6_w1500.jpg

Salesforce.com
https://www.salesforce.com/blog/2017/11/15-sales-statistics.html

The LUCK Principle by Geoff Ables
www.c5insight.com

Viral Leadership
https://i1.wp.com/richardnongard.com/wp-content/uploads/2018/09/Leadership-Infographic-and-Leadership-Statistics-Nongard.jpg?fit=1600%2C3461&ssl=1

Made in the USA
Columbia, SC
27 September 2022